Focus On
"SAP BW and ABAP"

Good Programming in SAP BW incl. HANA

©Matthias Zinke

Edition 2.0 "2016"

ISBN-13: 978-1530289967

ISBN-10: 1530289963

Contents

Your First Time with the Editor 6

Good Data Modelling 9

The Start Routine 12

Loop over the Data 16

Lookups 18

RECORDMODE in DSO 20

Coding Hints for This and That 22
    Variables in DTP 22

Start or End Routine? 24

Practical Examples 25
    Building a monthly average 25

All about Dates, Months, and Years 28
    Using the Right Format 28
    Using Offset to Obtain the Correct Information 28
    Retrieving the Last Day of a Month 29

All about Internal Tables 30
    Declaration of an Internal Table 30
    Standard, Sorted, or Hash Table 31
    Updating Internal Tables 32
    Reading Internal Tables 32
    Sorting and Deleting Internal Tables 33
    System Fields for Internal Tables 33
    Another Function in Internal Tables 34

Generic ABAP 35
    One Select Statement for More Than One Table 35
    Calling a Function Module Generic 35

Working with the ASSIGN COMPONENT OF STRUCTURE Statement 36

Performance Analysis and Debugging 37
Database vs. ABAP Runtime 38

ABAP and HANA 39

All about SELECT in SAP BW for HANA 41

Good Naming 42

Using Database HINT for HANA 43

The Author 45

Appendix 46
Table of Figures 46
Table of Coding 46
Index 47

How To

How have I come to write a book about my daily work? During the past 14 years of my SAP BW work life, I have met many developers, data warehouse architects, and newbies. During this time, I have seen thousands of lines of code in ETL and other BW areas. Fascinated by different ways of programming, I created my own ABAP Guideline, especially for handling millions of records in just a few minutes. The output or even better a private lesson that I learned is found here in this book on the following pages. If you find yourself working in an area of performant programming and even if you know a lot about good style coding, you can always take the opportunity to re-think your meanings given another view on ABAP. You will find the key to open the door of good programming practice in the pages of this book.

You can read this book from the beginning to end—my preferred way to do it—and later use it as a manual for daily reference.

Headlines provide the focus of each chapter. Do not worry about titles; you will always find answers to questions raised. Moreover, the structure of this book may not be as per your expectations compared with the traditional IT Guideline; my intent here is to rouse your interest rather than weigh you down with hundreds of pages of information that is available for free on the Internet.

If you have additional topics or details from your own desk, please send me an email at zinke.matthias@gmx.de and I will update this e-book with your topic. My best regards to you for this in advance.

***

Good Programming in SAP BW using ABAP is my personal view on how to set up coding for managing millions of records in the most performant manner. Maybe, you will be surprised that there are not 500 pages devoted to the secrets of performant ABAP coding but that I have covered the main points of interest in less than 50 printed pages. Thus, you can start reading without wading through an endless theoretical discourse about ABAP. If you have prior experience with BW and ABAP, you will find many hints for improving your coding. If you are new to this topic, after reading this book, you can begin coding without falling into simple beginner traps.

Berlin, Germany, April 2016

***

Trademarks

SAP and other SAP products and services mentioned herein as well as their respective logos are trademarks or registered trademarks of SAP SE (or an SAP affiliate company) in Germany and other countries.

All other product and service names mentioned are the trademarks of their respective companies. The pictures are taken from a SAP BW 7.4 Version.

Disclaimer

All information has been well prepared and tested, but the use of any content from this document is at your own risk.

***

Your First Time with the Editor

Transaction Code SE80

Please skip this brief introduction if you have prior experience with the editor.

If you want to use the entire transaction SE80 ABAP Workbench, you have to request for a developer key from your SAP Basis Team for the development system.

If you first create an ABAP report or an INCLUDE, you must enter this key one time in the prompt.

Let people know what you are doing! In the Attributes of a report or INCLUDE, enter at least the following Attributes so that your colleagues know, in general, what the report is about (Figure 1).

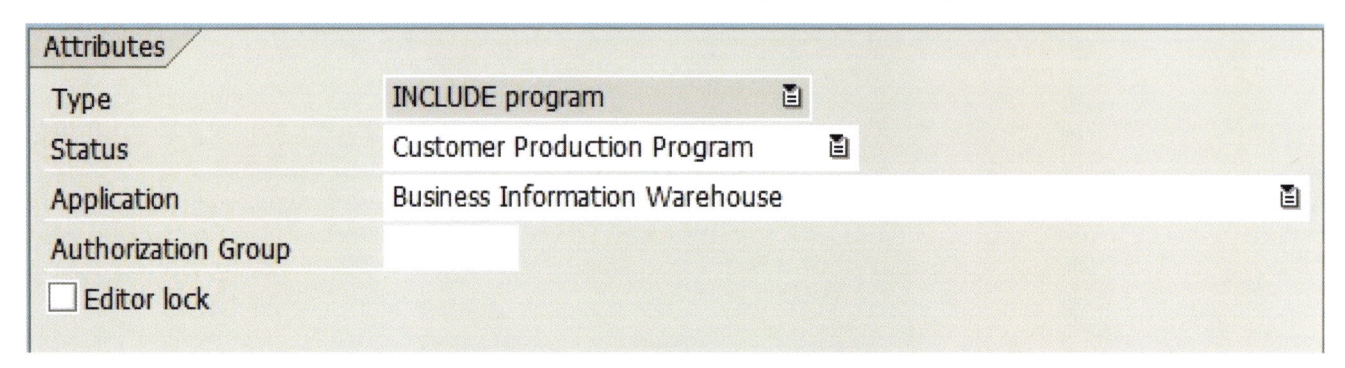

Figure 1 - Attributes of an ABAP Report

To make life easier for you and your colleagues, set the following Settings in the SE80, which you can find on the MENU bar at the top.

Use the new front-end editor, to make available the full power of the Workbench (Figure 2).

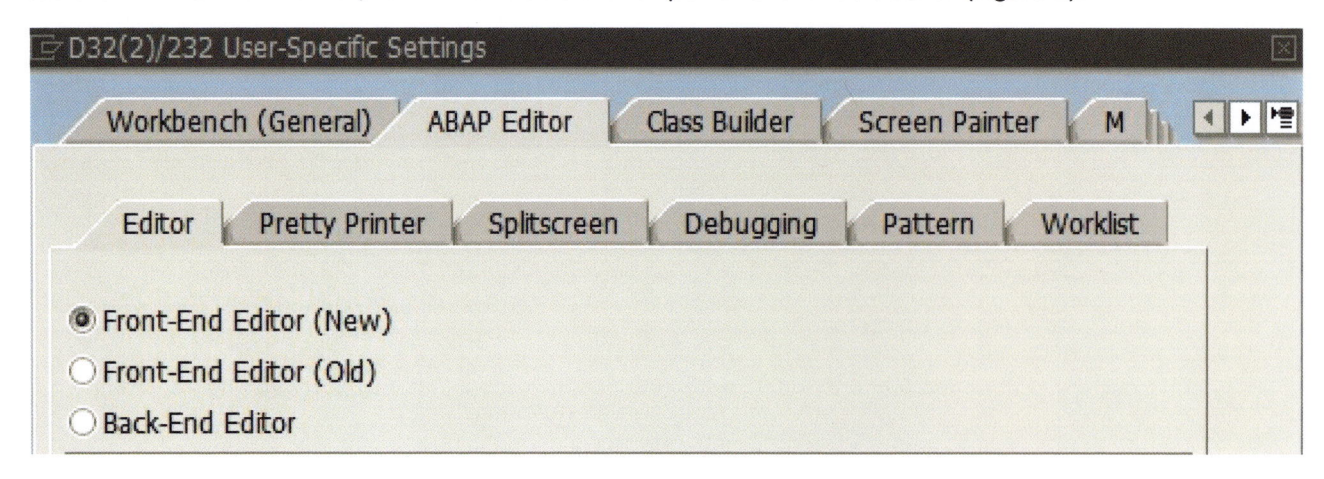

Figure 2 - User Specific Settings in the SE38

Use Pretty Printer to make the code more readable. My favorite setting in the Pretty Printer section is shown in Figure 3.

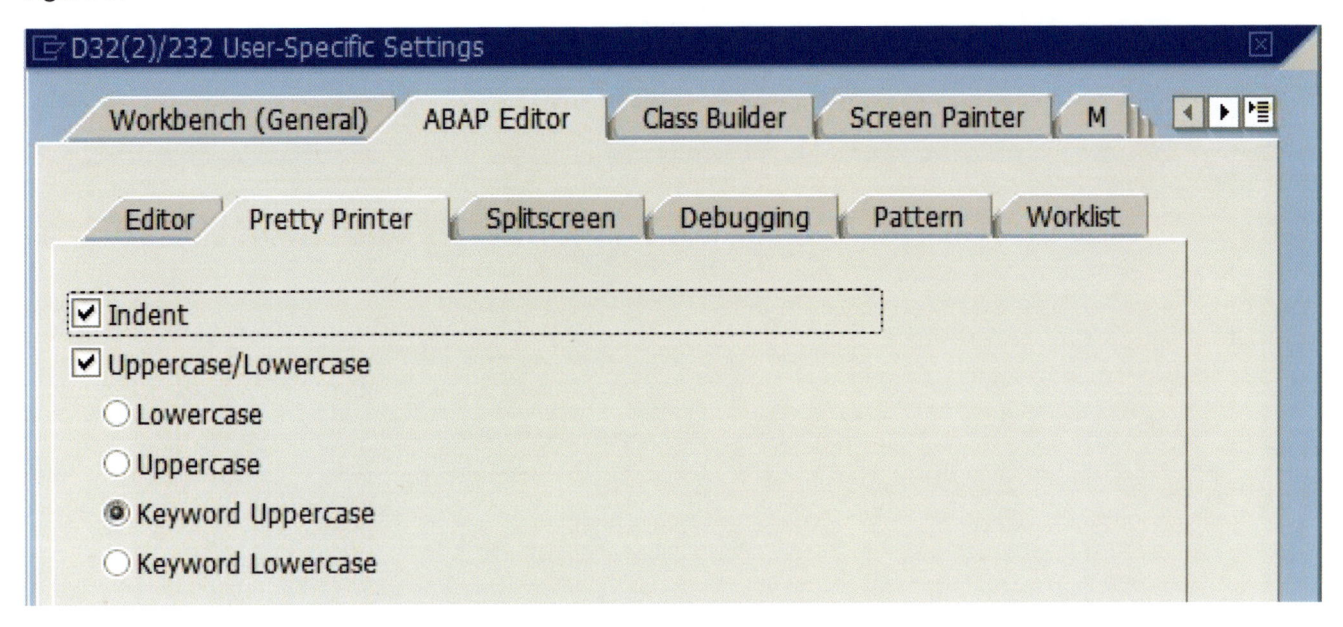

Figure 3 -Settings for the Pretty Printer

Another hint for making life easier is that, for **code completion,** you can use the following options, which are found in the lower right screen corner (Figure 4).

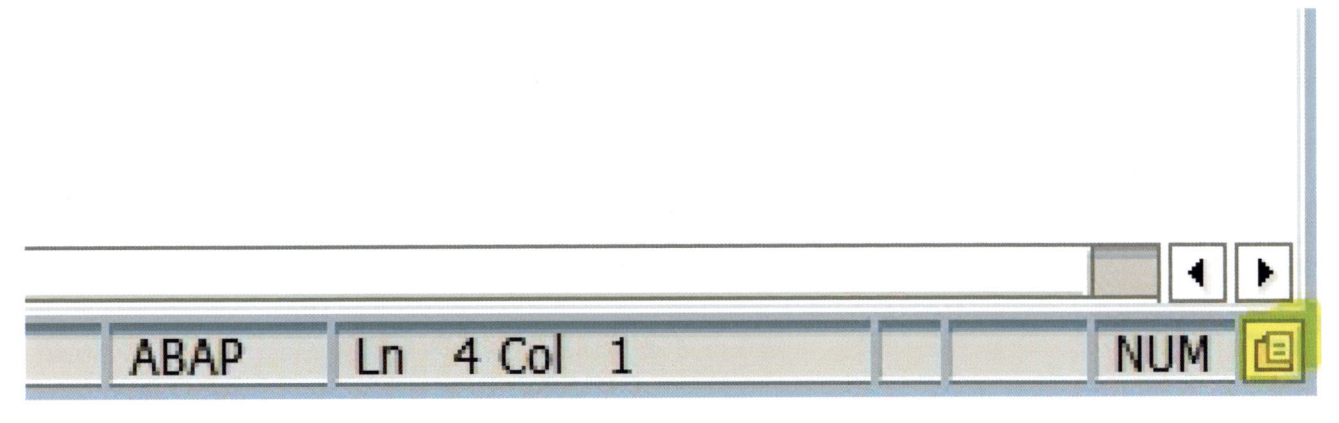

Figure 4 - Code Completion I

The option: **"Suggest Non-Keywords from the Text"** is necessary for the fast writing of a report in ABAP environment (Figure 5).

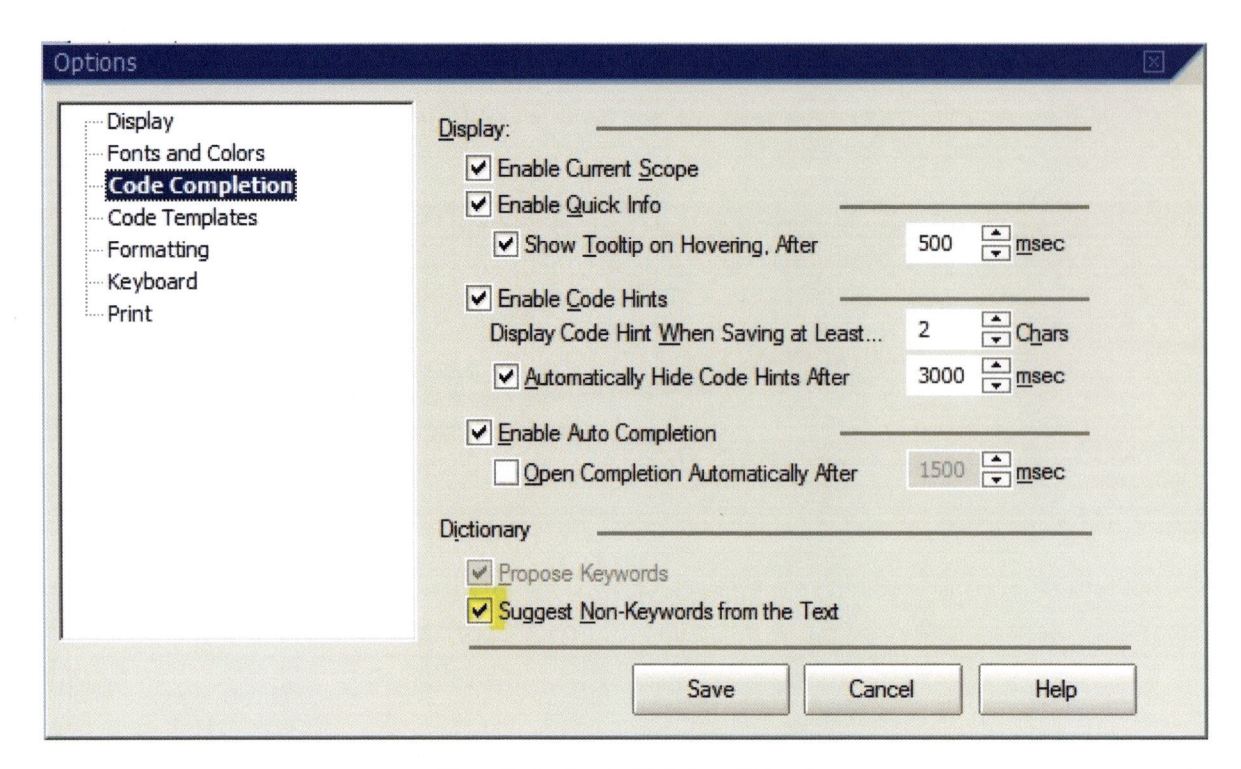

Figure 5 - Setting for Code Completion II

***

Good Data Modelling

As we are using HANA in this book, you can usually use DSO Objects as the common object for storing data in the SAP BW System.

To talk about good coding guidelines in SAP BW, you must also follow a simple little design guideline to access the full power of the SAP BW system.

If you create a data flow, e.g., from one DSO to another, you can use a transformation for to do so (Figure 6).

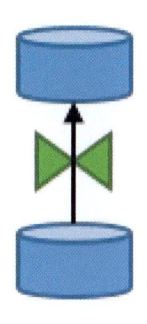

Figure 6 - Simple Data Flow

A better way to do this is to create two Info-Sources between them. Why? Doing so uses the full power of both the start and end routines, as shown in Figure 7.

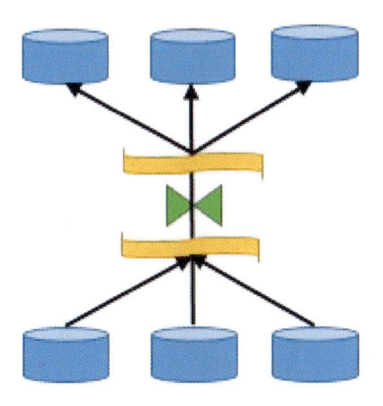

Figure 7 - Data Flow with Inbound- and Outbound Info-Source

The important fact to remember for these Info-Sources is that you should always use the same fields (Info Objects) in both.

You can easily create these Info-Sources if you are using the table RSDODSOIOBJ to identify the fields of the used DSO Objects (Figure 8, Figure 9 , and Figure 10).

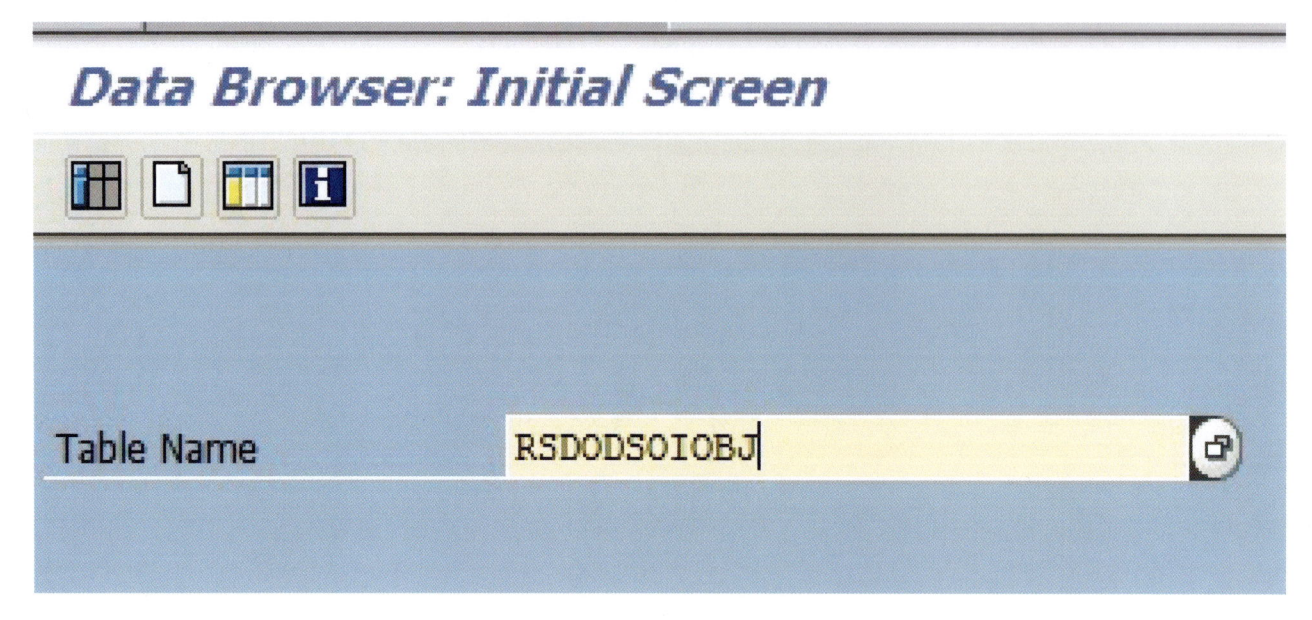

Figure 8 - Table RSDODSOIOBJ

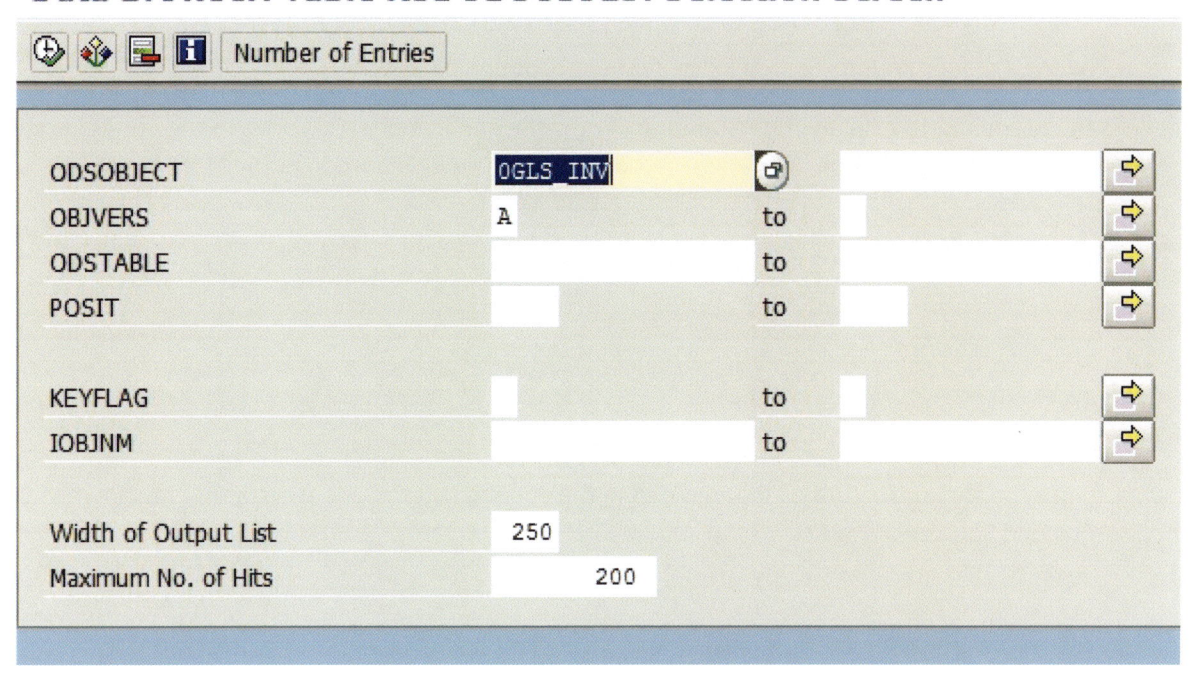

Figure 9 – Example for SE16 and Table RSDODSOIOBJ

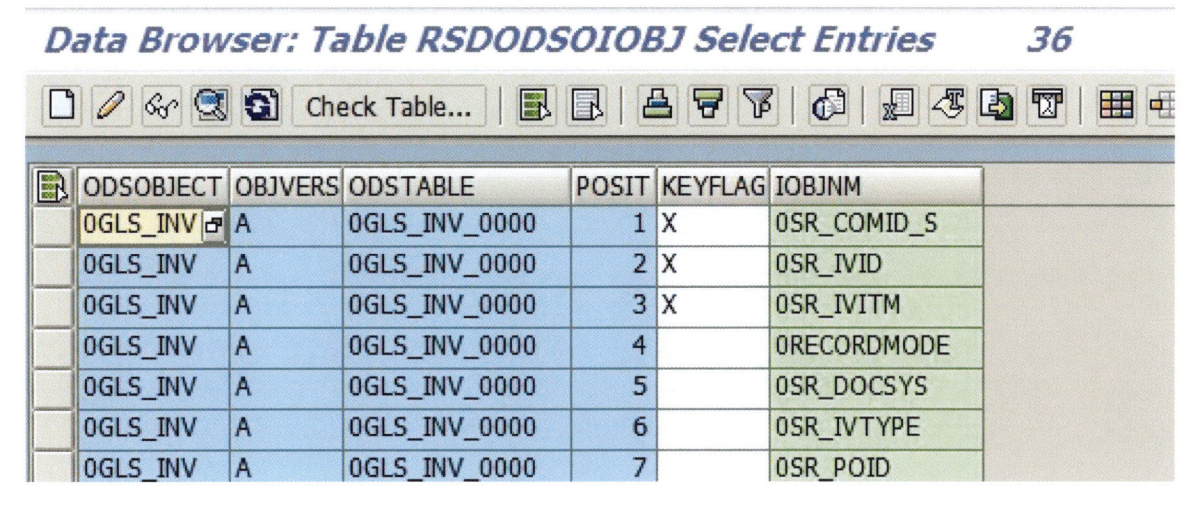

Figure 10 – Output of Transaction SE16

Now, you can copy the Info Object names to your clipboard or Excel spreadsheet and combine them with the Info Object of the target or source DSO. From there, you can insert the names in the Info-Source Editor simply using copy and paste. Why should you not use the "create with template button" to create the transformation? If you do, you do not get all the fields from the source and the target object into the Info-Source.

***

The Start Routine

Now we can address the programming step in the first start routine of this book.

Look at "predefine coding" to gain an understanding of how the generated part works. In the pre-generated coding, you will find the main section, as shown in Coding 1 - The Type-Definition in the Start-Routine.

```
*     Rule specific types
      TYPES:
         BEGIN OF _ty_s_sc_1,
```

Coding 1 - The Type-Definition in the Start-Routine

If you want to use the same structure as that defined in the coding, you can use the statement shown in Coding 2 - Type for a second Source Package.

```
DATA: ls_source_package TYPE _ty_s_sc_1.
```

Coding 2 - Type for a second Source Package

Even if there are changes in the structure (Info-Source), you do not have to adapt your coding. Now is a good time to explain how to type used objects in the start routine and how to name them (Coding 3).

```
DATA: ls_source_package TYPE _ty_s_sc_1. "Local Structure

DATA: lt_source_package TYPE TABLE OF _ty_s_sc_1. "Local Table

DATA: lt_source_package TYPE _ty_t_sc_1. "Local Table with table type

DATA: lv_date TYPE d. "Local Variable

FIELD-SYMBOLS: <new_dp> TYPE _ty_s_sc_1. "Local Field Symbol
```

Coding 3 - Naming proposal for Local Objects

For the source package, the type definition starts at this point (Coding 4).

```
TYPES:
    _ty_t_SC_1          TYPE STANDARD TABLE OF _ty_s_sc_1
                        WITH NON-UNIQUE DEFAULT KEY
```

Coding 4– Definition of the Structure for the Source Package

The source package itself has reverted to this type (Coding 5).

```
CHANGING
    source_package                  type _ty_t_sc_1
```

Coding 5 - Definition (Type) of the Source Package

The other parameters used by the start routine revert to standard dictionary definitions, which you can use for your coding (Coding 6).

```
METHODS
    start_routine
      IMPORTING
        request                 type rsrequest
        datapackid              type rsdatapid
        segid                   type rsbk_segid
      EXPORTING
        monitor                 type rstr_ty_t_monitors
      CHANGING
        SOURCE_PACKAGE          type _ty_t_SC_1
      RAISING
        cx_rsrout_abort
        cx_rsbk_errorcount.
```

Coding 6 - The Standard Type Definition in the Start Routine

For example, the monitor table type (Coding 7 and Figure 11) is as follows:

```
rstr_ty_t_monitors    TYPE STANDARD TABLE OF rstmonitor
                      WITH NON-UNIQUE DEFAULT KEY,
```

Coding 7 – The Type for the Monitor Table

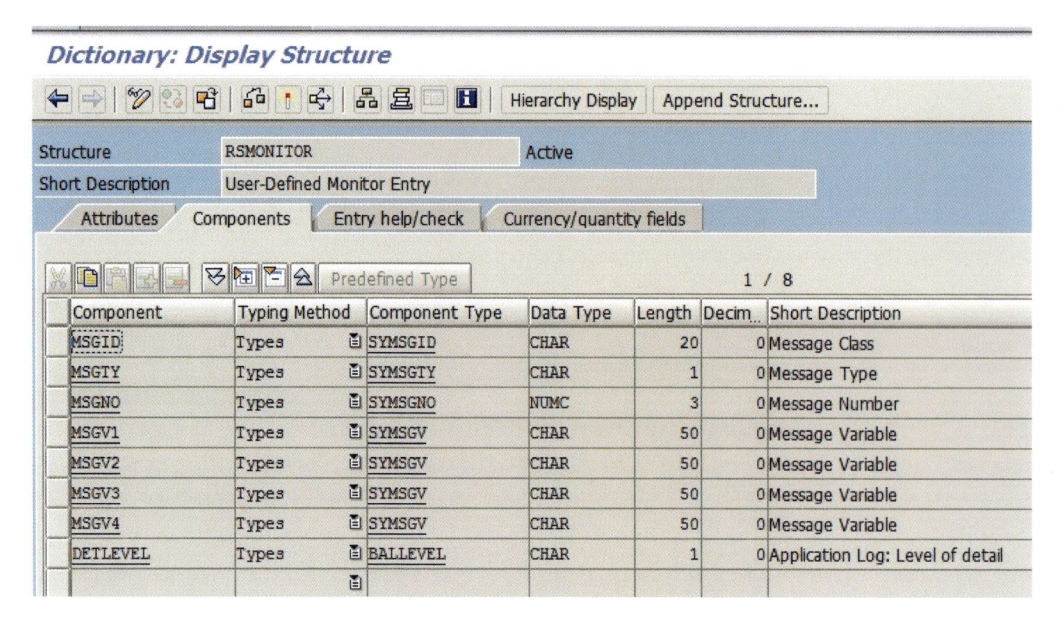

Figure 11 - The Structure Definition of the Monitor Table

The first area where you can enter your coding is as follows (Coding 8).

```
*$*$ begin of global - insert your declaration only below this line   *-*
    ..."insert your code here
*$*$ end of global - insert your declaration only before this line   *-*
```

Coding 8– Begin of the Global Part for the Start Routine

Please remain seated! Be very careful with this section, as you must understand that this is a global declaration, and of course, the content of this part remains alive throughout the run time of the transformation. If you forget, for example, to clear the defined variables and tables at this point, they will be grown by running more than one Info-Package.

The next step in free coding is as follows (Coding 9).

```
*$*$ begin of 2nd part global - insert your code only below this line  *
... "insert your code here
*$*$ end of 2nd part global - insert your code only before this line  *
```

Coding 9 – The Second part of the Global Part in the Start Routine

Here you can enter, for example, a select statement and other tasks you want to complete only one time in the transformation.

The most important part is the following section (Coding 10).

```
*$*$ begin of routine - insert your code only below this line        *-*
    ... "insert your code here
*--  fill table "MONITOR" with values of structure "MONITOR_REC"
*-   to make monitor entries
    ... "to cancel the update process
*    raise exception type CX_RSROUT_ABORT.
```

Coding 10 – The Main Part for the Coding in the Start Routine

Now, we can finally start to enter our own start routine.

> *Please note that the start and end routines are not fully integrated with the version management of SAP, which means that if you delete one of the routines by mistake, there is no way to recover the coding.*

To avoid any trouble caused by the lack of version management, you can always create an INCLUDE in the SE80 transaction and enter this in the Start Routine. The advantage of INCLUDE is that it will be used as a normal part of the coding, as if you had entered the coding directly into the start routine.

To call INCLUDE, you must use the INCLUDE statement in your coding (Coding 11).

```
INCLUDE zbw_start_dso1_dso2.
```

Coding 11 – Using INCLUDE in the Start Routine

Now the source is saved in the database with full version management behind it.

***

Loop over the Data

If you want to change the data in the routine, you must loop over the source package. There are two ways to do so. The first is the performant way, but if you choose this option, the changes you make in the coding will directly change the source package (Coding 12).

```
LOOP AT source_package ASSIGNING <source_fields>.

  ADD 1 TO <source_fields>-calday.

ENDLOOP.
```

Coding 12 – Ssimple LOOP over the Source Package

A simple example is to change the calendar day and always add one day to the given date.

The second possibility is to use the internal structure, also called the work area, to change one simple line of the data. To do so, you must use the following coding instead (Coding 13).

```
LOOP AT source_package INTO ls_source_package.

  ADD 1 TO ls_source_package-calday.

  APPEND ls_source_package TO lt_new_dp.

ENDLOOP.

FREE source_package.
source_package[] = lt_new_dp[].
FREE lt_new_dp.
```

Coding 13 – Using work-areas instead of pointers

As shown in the example in Coding 13, the second method requires a lot of more coding to achieve the same result. However, sometimes it is necessary to do so. If you are using this coding method, do not forget to clear or free the used objects prior to doing so (Coding 14).

```
FREE lt_new_dp.
CLEAR ls_source_package.
```

Coding 14 – FREE and CLEAR Objects After or Before Using Them

Let's briefly review the advantages of these two possible options. The loop with the ASSIGNING statement is the more performant way to make the change. The reason is that no real object is used to temporally store the data; we are simply using a pointer to go over the source package. The changes are then made directly in the source package.

With the second method, we are using real objects that we change and that takes more time in the run-time environment. The advantage of this option is that we can use the old values in the source package for the next loop.

***

Lookups

The main reason to introduce a start routine is to initiate lookups for data stored in other DSO or Info-Objects.

As such, we want to initiate these lookups in the most performant way.

The most important statements in lookups are the SELECT and READ statements.

First, we select the data from the other DSO Object. Therefore, we need at least two objects to store these data in our coding.

The data are stored in the data base table, as shown below (Figure 12).

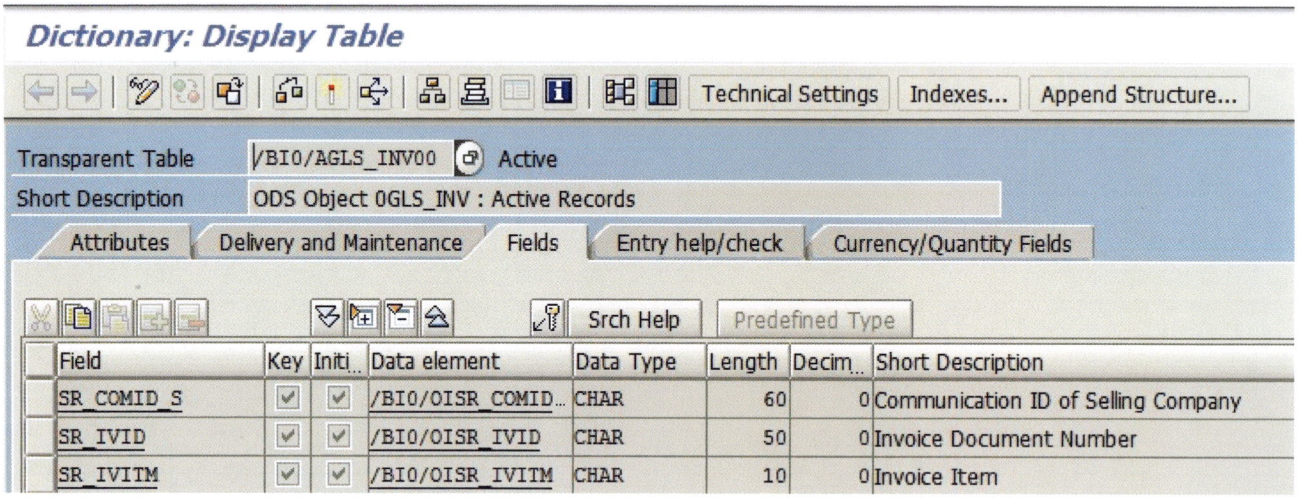

Figure 12 – Example for a Look-up Table (DSO)

We can use the name to identify the internal table and the structure used (Coding 15).

```
DATA: lt_gls_inv TYPE TABLE OF /bi0/agls_inv00.
DATA: ls_gls_inv TYPE /bi0/agls_inv00.
```

Coding 15 – The Correct Type for the local objects

To select all the data we want to use in our LOOP over the data, it is a good idea to select them all at once in the beginning of the routine. To do so, we can use the FOR ALL ENTRIES IN addition (Coding 16).

```
SELECT * FROM /bi0/agls_inv00 INTO TABLE lt_gls_inv
  FOR ALL ENTRIES IN source_package
  WHERE calday = source_package-calday.
```

Coding 16 – Select Statement for a Lookup from a DSO

Prior to using FOR ALL ENTRIES IN, always check that the internal table used is not empty; otherwise you will get back all the data from the database, which may be a lot (Coding 17).

```abap
IF source_package IS NOT INITIAL.

  SELECT * FROM /bi0/agls_inv00 INTO TABLE lt_gls_inv
    FOR ALL ENTRIES IN source_package
    WHERE calday = source_package-calday.
ENDIF.
```

Coding 17 - Checking the Content before using FOR ALL ENTRIES

At this point, we have made a single fetch to select data from the database. The next step is to read those data during our LOOP over the package (Coding 18).

```abap
LOOP AT source_package ASSIGNING <source_fields>.

  CLEAR ls_gls_inv.

  READ TABLE lt_gls_inv INTO ls_gls_inv
  WITH KEY calday = <source_fields>-calday.

  IF sy-subrc = 0.
    <source_fields>-sr_conid = ls_gls_inv-sr_conid. "Local Contract ID
  ENDIF.
ENDLOOP.
```

Coding 18 – The READ TABLE Statement in a LOOP

In the chapter **All about Internal Tables,** you will find explanations regarding the use of additions to the READ Statement or the use of other types of internal tables.

***

RECORDMODE in DSO

The real property of the 0RECORDMODE is to guarantee the setup delta for DSO Objects. If you want to manipulate this behavior, you must create, for example, a self-update transformation for a DSO Object (Figure 13, - Self-update for a DSO Object).

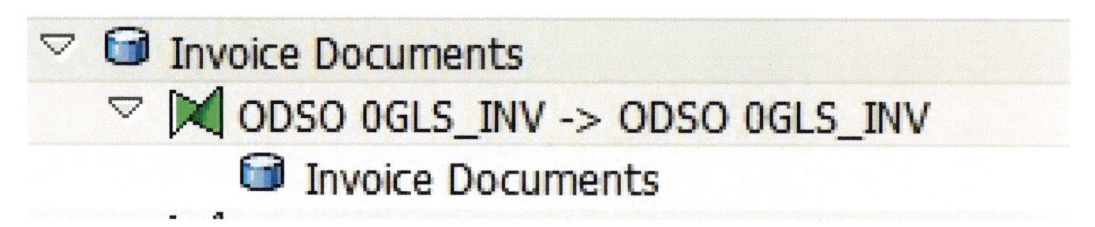

Figure 13 - Self-update for a DSO Object

Mapping of the included fields can be in a 1:1 ratio, or whatever you require for the self-update (Figure 14).

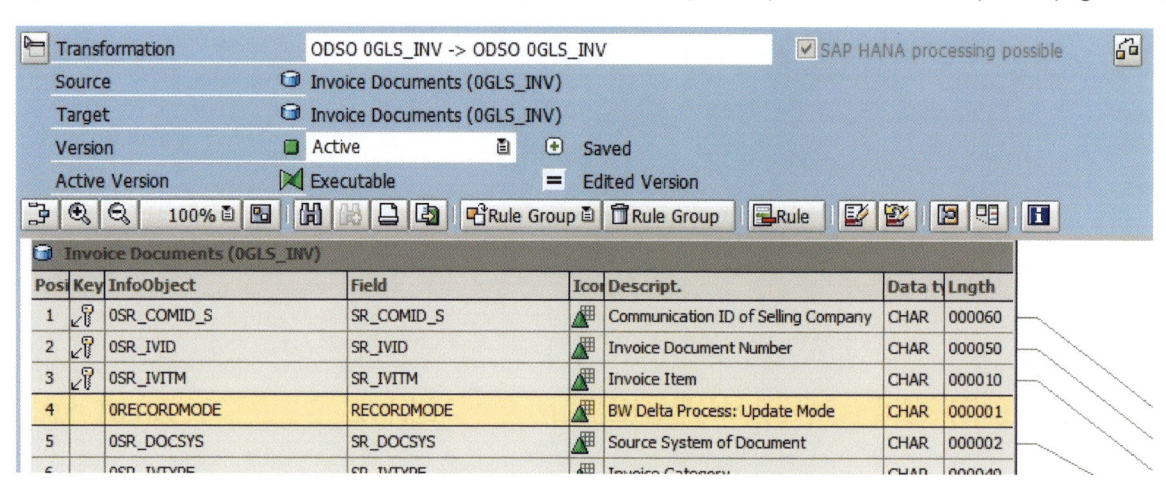

Pos	Key	InfoObject	Field	Icon	Descript.	Data t	Lngth
1	🔑	0SR_COMID_S	SR_COMID_S		Communication ID of Selling Company	CHAR	000060
2	🔑	0SR_IVID	SR_IVID		Invoice Document Number	CHAR	000050
3	🔑	0SR_IVITM	SR_IVITM		Invoice Item	CHAR	000010
4		0RECORDMODE	RECORDMODE		BW Delta Process: Update Mode	CHAR	000001
5		0SR_DOCSYS	SR_DOCSYS		Source System of Document	CHAR	000002
6		0SR_IVTYPE	SR_IVTYPE		Invoice Category	CHAR	000040

Figure 14 - 1:1 Mapping of the Fields

As shown in - 1:1 Mapping of the Fields, the 0RECORDMODE is not linked to any target object. So, you will have to go to the Technical Rule Group to see the target 0RECORDMODE object (Figure 15).

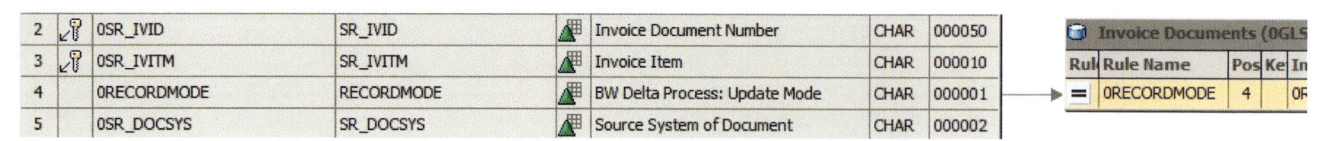

Figure 15 - Technical Rule Group for the 0RECORDMODE

To manipulate the data in the simplest way, create a start routine with coding such as that shown in Coding 19

```
LOOP AT source_package ASSIGNING <source_fields>
  WHERE calday = sy-datum.

  <source_fields>-recordmode = 'D'.  "Delete this record

ENDLOOP.
```

Coding 19 – Manipulating the 0RECORDMODE in a Start Routine

The above code deletes all records that are less than or equal to January 1, 2015. Of course, this can also be done via the selective deletion option in the DSO, but you can imagine that you can easily create a DTP and then execute it via a process-chain. When using the select option in a DTP, you can delete every record you wish in the DSO.

Another approach is do it the other way around; you can create records via a self-update and the coding could then follow. However, ensure that you change the key of the DSO or you will overwrite the existing data via the routine (update or activation of the DSO) (Coding 20).

```
LOOP AT source_package ASSIGNING <source_fields>
  WHERE calday = sy-datum.

  <source_fields>-sr_conid = <source_fields>-sr_conid + 1.
  <source_fields>-recordmode = 'A'.  "Add this record"

ENDLOOP.
```

Coding 20 – Adding records in a Start Routine of a DSO

If you only want to loop over the data in the DSO, you needn't change the 0RECORDMODE at all, because the 0RECORDMODE will be automatically set to "N," which means a new record and an update of the existing record.

***

Coding Hints for This and That

In this chapter I provide some coding crumbs for various tasks in the work of a BW developer.

Variables in DTP

First, we'll look at the DTP variables, where you can create some coding in the selection area of the filter value.

For example, you should always set the date and full load to the actual month. However, you do not have the 0CALMONTH in the source. So, you must figure out which days in the routine belong to the actual or given month.

The coding can be set up as follows (Figure 16).

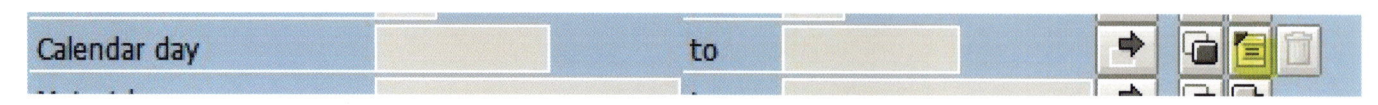

Figure 16 - Routines in DTP Variables

```
DATA: l_idx LIKE sy-tabix.
DATA: lv_date TYPE d.
DATA: lv_first_day TYPE d.

READ TABLE l_t_range WITH KEY
     fieldname = 'CALDAY'.
l_idx = sy-tabix.

lv_date = sy-datum.
SUBTRACT 1 FROM lv_date.    "Load from yesterday
lv_first_day = lv_date(6).
lv_first_day+6(2) = '01'.  "The first day of the actual month

l_t_range-iobjnm = 'OCALDAY'.
l_t_range-fieldname = 'CALDAY'.
l_t_range-sign = 'I'.
l_t_range-option = 'BT'.
l_t_range-low = lv_first_day.
l_t_range-high = lv_date.

**....
  IF l_idx <> 0.
    MODIFY l_t_range INDEX l_idx.
  ELSE.
    APPEND l_t_range.
  ENDIF.
  p_subrc = 0.
```

Coding 21 – Example for Date Selection

As a result, you will always get the current values for the actual month, excluding the actual day (Coding 21 and Figure 17).

Field	S	OP	From value	To value
CALDAY	I	EQ	20160201	20160215

Figure 17 - The Result of the Coding from Coding 21

***

Start or End Routine?

There is an easy answer to this question! The Start routine is used for changing the source fields and the End routine for changing the result fields of a transformation. However, if we consider that we use the same structure in the two Info-Sources around our transformation, then the source and result fields are equal.

Why do we then separate the code into Start and End routines? The best explanation is to give an example of the calculation of a monthly average for a given set of key figures, and in the same transformation working with the calculated average to make some additional changes or calculations.

If we are looking in the Detail View of a transformation, we can see that there are always two steps for each transformation. You can switch on the Detail View by using this button.

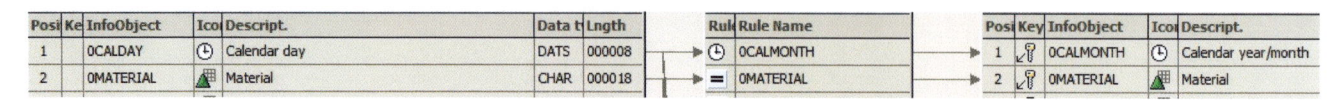

Posi	Key	InfoObject	Icon	Descript.	Data t	Lngth		Rule	Rule Name		Posi	Key	InfoObject	Icon	Descript.
1		0CALDAY		Calendar day	DATS	000008	→	⊕	0CALMONTH	→	1	🔑	0CALMONTH		Calendar year/month
2		0MATERIAL		Material	CHAR	000018	→	=	0MATERIAL	→	2	🔑	0MATERIAL		Material

Figure 18 - Detail View of a Transformation

As you can see in Figure 18, the transformation is divided into three steps:

1. Start routine
2. Field rules for every single Info-Object
3. End routine

As a result, you can use the Info-Source as an additional layer to make a data aggregation on the way from one DSO to another.

***

Practical Examples

Building a monthly average

The example of how to set up a monthly average is mentioned in the above chapter "Start or End Routine?" and here I explain this in detail.

The task for setting up a monthly average is as follows:

"A set of given key figures is related to the calendar day stored in 0CALDAY. We want to set up a monthly average which, in the result, is related to the calendar month 0CALMONTH. In the second step we want to calculate a new key figure based on the average."

One possible solution is to build two new DSO Objects, one with the aggregation function (changing the key of the DSO from 0CALDAY to 0CALMONTH) and the second for the calculated new key-figures. This is a poor solution if we consider the space consumption of our DSO objects, since it requires storing the same set of data twice in our BW system.

The better solution is to use the concept of central transformation, with two Info-Sources between the given set of data and the target DSO, with the calculated average and the newly calculated key figure.

The keys of the Info-Sources are then the new aggregation, including 0CALMONTH, and the mode of the aggregation behavior must be setup as a summation (ignore the warning about this summation aggregation during the activation of the Info-Source).

The coding to calculate the average is divided into two sections:

1. Calculate the number of days for a given month based on the given calendar day.
2. Calculate the average for all key figures in a compressed way.

To do the first step, see the coding in Coding 22

```
CALL FUNCTION 'SLS_MISC_GET_LAST_DAY_OF_MONTH'
        EXPORTING
           day_in              = <source_fields>-calday
        IMPORTING
           last_day_of_month   = lv_last_date
        EXCEPTIONS
           day_in_not_valid    = 1
           OTHERS              = 2.
     IF sy-subrc <> 0.
*  Implement suitable error handling here
     ENDIF.
     lv_daycount = lv_last_date+6(2).

     ELSEIF  <source_fields>-calday(6) = lv_yesterday(6).
       "For the current month
       lv_last_date = lv_yesterday.
       lv_daycount =  lv_last_date+6(2).

     ENDIF.

     <source_fields>-calday = lv_last_date.
```

Coding 22 – Calculation of Number of Days for a given Calendar Day

The coding uses a standard function module to retrieve the last day of a month from a given day. Then, we set the variable lv_daycount with an offset to the correct numbers of days. For the current month, this calculation is not correct. Maybe we are in the middle of the month and have only 15 days to calculate the average for the current period. The last statement, for quality assurance, is to store the last day of the month in the source field 0CALDAY.

The second step belongs to the chapter Generic ABAP, where I will explain in detail the coding shown in Coding 23.

```
LOOP AT lt_rsksfieldnew ASSIGNING <rsksfieldnew>.
   "Sructure with all key-figures

   ASSIGN COMPONENT <rsksfieldnew>-fieldnm OF STRUCTURE
   <source_fields> TO <keyfigures>.

      <keyfigures> = <keyfigures> / lv_daycount.
ENDLOOP.
```

Coding 23 – Second Step: Calculation of the Average using ASSIGN COMPONENT

The structure with all the key figures was pre-selected during the global part of the transformation. The source for the fields is the Inbound Info-Source from our central transformation (see Coding 24).

```abap
SELECT * FROM rsksfieldnew INTO TABLE lt_rsksfieldnew
    WHERE isource = 'IS_DSO1_O' "Outbound InfoSource
    AND objvers = 'A'
    AND unifieldnm = '0BASE_UOM' "Only QTY
    AND type = 'KYF'. "All Keyfigures
```

Coding 24 - Select statement from table RSKSFIELDNEW

Here, the table RSKSFIELDNEW is the SAP helper table where all data are stored for the new Info-Source, i.e., new since BW 7.0

***

All about Dates, Months, and Years

This chapter addresses the use of the date, month, and year functions and fields in SAP BW.

Using the Right Format

If you are working with the dates, months, and years, always use the correct format for these fields (Coding 25), and also in table definition (Figure 19).

```
DATA: lv_date TYPE d.
DATA: lv_year(4) TYPE n.
DATA: lv_month(2) TYPE n.
```

Coding 25 – The right Format for the Date in ABAP Coding

Field	Data element	Data Type	Length	Decim...	Short Description
CALDAY	/BI0/OICALDAY	DATS	8	0	Calendar day
CALYEAR	/BI0/OICALYEAR	NUMC	4	0	Calendar year
CALMONTH2	/BI0/OICALMONTH2	NUMC	2	0	Calendar month
CALMONTH	/BI0/OICALMONTH	NUMC	6	0	Calendar year/month

Figure 19 - Correct Format to Use in a Dictionary Definition.

Internally, during the run time and at the database level, the format for the date is always YYYYMMDD. For example, the last day this year is 20161231.

Using Offset to Obtain the Correct Information

To manipulate the date field, you can use the offset technique to make the change you want (Coding 26).

```
lv_date(4)   = sy-datum(4).    "The year of the date
lv_date+4(2) = sy-datum+4(2).  "The month of the date
lv_date+6(2) = sy-datum+6(2).  "The day of the date
```

Coding 26 – Using Offset for the date Field

Retrieving the Last Day of a Month

How do you retrieve the last day of a given month? If you are using date type in your coding, then doing so is easy, because the SAP internal function of this "DATS" fields now relates to the calendar.

```abap
DATA: lv_last_date TYPE d.

lv_last_date = sy-datum.

lv_last_date+4(2) = lv_last_date+4(2) + 1.

IF lv_last_date+4(2) = '13'. "In December
  lv_last_date(4) = lv_last_date(4) + 1. "Next Year
  lv_last_date+4(2) = '01'. "January

ENDIF.

lv_last_date+6(2) = '01'.

SUBTRACT 1 FROM lv_last_date.
```

Coding 27 – Using SUBTRACT 1 to Retrieve the Last Day of a Month

***

All about Internal Tables

All you need to know about internal tables is described in this chapter. In general, internal tables are run-time objects, which means that they remain alive only during the execution of your code. The second important thing to know is that they are consuming or, at best, allocating memory in the application server in the RAM area of your hardware. That has led to the general assumption that you must keep these objects as small as possible. In addition, if and when you do not need them anymore, delete them out of memory with the statement shown in (Coding 28).

```
FREE lt_table.
```

Coding 28 – Cleaning out Internal Tables

DO NOT use the REFRESH Statement shown in Coding 43, because it has been marked as obsolete.

```
REFRESH lt_table.
```

Coding 29 – DO NOT USE the Obsolete REFRESH Statement

Declaration of an Internal Table

To declare an internal table you must first structure the content of this table. The easiest and usually the best way to do so is to copy a structure definition from the SAP dictionary (Coding 30).

```
DATA: lt_gls_inv TYPE TABLE OF /bi0/agls_inv00.
```

Coding 30 – Table Reference to the Dictionary

Use this type of declaration only if you are sure you want to use the complete structure of the dictionary table, or at least more than half of it, and you want to have the internal table dynamically adjusted as the dictionary changes.

If you have to reduce the allocation of memory, you can also define within your coding a local table type for the table definition (Coding 31).

```
TYPES: BEGIN OF lty_table,
       calday TYPE /bi0/oicalday,
       END OF lty_table.
```

Coding 31 – Local Definition of a Table Type

Afterwards, you can use this type for the declaration of your internal table (Coding 32).

```
DATA: lt_table TYPE TABLE OF lty_table.
```

Coding 32 – Table Definition with a Local Type

When I do so, I also always declare a structure (work area) with the same type, as you will need this in 90% of your coding work (Coding 33).

```abap
DATA: ls_table TYPE lty_table.
```

Coding 33 – Structure Definition with the same Local Type as in Coding 32

Enhancement of Dictionary Tables

Sometimes, you need only one additional field for an internal table. Should you declare the whole type again? The answer is No. You can also enhance the structure of your internal table with the addition of a field or another structure.

Please be careful with this issue, read the SAP Help regarding the INCLUDE - TYPE, STRUCTURE Statement before doing this, and rename your component.

Standard, Sorted, or Hash Table

For the READ statement used later, you must consider the key and the sort order of your internal table. In the declaration section, you can use the following additions (Coding 34).

```abap
DATA: lt_table TYPE TABLE OF lty_table WITH KEY calday.
DATA: lt_hash_table TYPE HASHED TABLE OF lty_table WITH UNIQUE KEY calday.
DATA: lt_sort_table TYPE SORTED TABLE OF lty_table WITH UNIQUE KEY calday.
```

Coding 34 – Standard, Sorted and Hash Tables

As you can see, there are differences between the three tables. The first is the standard table for which you can also define a key in addition to your declaration (optional). For the sorted table and the special sorted hash table, you must use the addition of a unique key (mandatory). The memory consumption for all three objects is the same. However, the advantage of a sorted table or a hash addition is the performance during the READ step. For internal tables with up to 10,000 record lines you don't really need this addition, but if you are handling larger objects in your ABAP, you will have to consider using this addition, and, if possible, using the hash table. The HASH table definition tells the READ statement how to read the table lines. In the end, the access time for reading a single line is the same every time for the HASH algorithm being used. I use this addition only if I discover that there is a bottleneck in my coding from the processing of the data. If you use this addition, you must insert the correct lines and sort order of those tables.

Updating Internal Tables

If you want to use an internal table as a buffer for new records in your transformation, you will have to update the records. One option you can use is shown in Coding 35.

```
APPEND ls_source_package TO source_package.
```

Coding 35 – Append of an Internal Table

If you want to move the entire content of an internal table to another internal table, you can use the coding shown in Coding 36. This is sometimes necessary if you want add data and re-use the whole source or result package.

```
source_package = lt_new_dp.
source_package[] = lt_new_dp[].  "For tables with header lines
```

Coding 36 – Completely Moving Content from one Internal Table to Another Internal Table

Reading Internal Tables

To find the right data in the internal table, use the READ Statement or loop over the internal table.

```
READ TABLE lt_gls_inv INTO ls_gls_inv
     WITH KEY calday = <source_fields>-calday.
```

Coding 37 – READ Statement with the addition KEY only

Be careful here with the automatic code completion if you are typing WITH in the code. The completion always suggests the WITH TABLE KEY. If you use this, remember you have to use the table key for the READ statement.

The second way to find data is a little bit slower but sometimes you might need it to read the correct data. For example, you may want to use the Operator LE (less or equal) to select a date range when reading the data. This is not supported with the READ Statement, so you have to use a simple LOOP to fetch the data (Coding 38).

```
LOOP AT source_package ASSIGNING <source_fields>
   WHERE calday LE sy-datum.

   EXIT.

ENDLOOP.
```

Coding 38 - LOOP and EXIT for getting data out of an Internal Table

The EXIT Statement indicates that you are leaving the LOOP after you have found the first set of data that matches the WHERE clause in the LOOP.

Sorting and Deleting Internal Tables

If you are using the non-sorted standard table in your coding and want to use a fast READ for this, you have to sort the table before you READ it. After sorting, you can also use the addition BINARY to optimize the READ statement (Coding 39).

```
SORT source_package BY calday.

READ TABLE source_package INTO ls_source_package
WITH KEY calday = sy-datum
BINARY SEARCH.
```

Coding 39 - Correct use of SORT and BINARY READ

Please refer to Figure 20:

The addition **BINARY SEARCH** produces a binary search of the table, not linear. In the case of large tables (from approximately 100 entries), this can significantly reduce runtime. The table must, however, be sorted in ascending order by the components specified in the search key. The priority of the sort order must match exactly the order of the components in the search key. If this requirement is not met, the correct row is not usually found.

Figure 20 SAP Help for BINARY SEARCH [1]

If you want to eliminate a couple of lines of an internal table w/o looping over it, you can use the DELETE statement (Coding 40):

```
DELETE source_package WHERE calday = sy-datum.
```

Coding 40 - DELETE Statement for an Internal Table

System Fields for Internal Tables

If you are using READ for an internal table, you can use sy-subrc to check whether READ found at least one record (Coding 41):

```
IF sy-subrc = 0.

ENDIF.
```

Coding 41 - Return Code of a Successful READ Statement

If you are looping over an internal table, you can check the lines of the actual loop using the sy-tabix field from ABAP runtime (Coding 41):

---

[1] From SAP HELP Page

```
    IF sy-tabix = 1.

    ENDIF.
```

Coding 42 - First Line of an Internal Table

Another Function in Internal Tables

If you want to count the number of lines in the actual internal table, use the in-build function (Coding 43):

```
    DATA: lv_lines TYPE i.

    DESCRIBE TABLE source_package LINES lv_lines.
```

Coding 43 – Describe Tables to retrieve the Number of Lines

***

Generic ABAP

The main advantage of using this technique to write generic ABAP coding is that you can re-use the same coding for more than one task and it is not necessary to know the exact name of objects when writing the ABAP. Of course, you must also follow some rules to avoid inadvertently generating a dump during the execution of your code.

One Select Statement for More Than One Table

The easiest way to understand this technique is to use variable instead of hardcode names. For example, you may want to extract data from two different sources (that have the same structure) without doubling the code.

```abap
DATA: lv_tabname TYPE tabname.
DATA: lt_dso TYPE TABLE OF /bi0/agls_inv00.

LOOP AT source_package ASSIGNING <source_fields>.

  IF <source_fields>-calday LE '20123112'.
    lv_tabname = '/BIC/ADSO100'.
  ELSE.
    lv_tabname = '/BIC/ADSO200'.
  ENDIF.

  SELECT * FROM (lv_tabname) INTO TABLE lt_dso.

ENDLOOP.
```

Coding 44 – Variable Selection from a Table

As shown in Coding 44, we define a variable for the table name such that during the runtime of the ABAP, we can automatically switch between the two source DSOs, which, at the very least, have date and time as common fields.

Calling a Function Module Generic

The same principle can be used to execute a function module. Let us assume that you want to make the exit for variables more flexible in your system. For this purpose, you can use the following approach (Coding 45):

```abap
DATA: lv_funvname TYPE rs38l_fnam.

CONCATENATE 'ZBW_' i_vnam INTO lv_funvname.

CALL FUNCTION lv_funvname
  IMPORTING
    i_vnam = i_vnam.
```

Coding 45 – Generic CALL of a Function module

Please note that for security reasons, some companies do not allow any generic calls. Nevertheless, the example in Coding 45 shows that you can create only one function module per variable, and then, the call of the automatic user exit is linked to your coding. You must check the existence of the function module to avoid dumps during the runtime; this is mandatory. Parameters must be defined just as the user exit function module EXIT_SAPLRRS0_001 is defined. The easiest way to create single-variable function modules is to copy those of EXIT_SAPLRRS0_001.

Working with the ASSIGN COMPONENT OF STRUCTURE Statement

The next step is to program a dynamic calculation based on a known structure. This very powerful statement links with the Component of the Structure to become a field symbol, whereby you can then directly change the content of the structure. Coding 46 shows how to implement an average calculation for all included quantities in some lines of coding.

```
ASSIGN COMPONENT <rsksfieldnew>-fieldnm OF STRUCTURE
<source_fields> TO <keyfigures>.
```

Coding 46 – Working with ASSIGN COMPONENT OF STRUCTURE TO Statement

If you are working with field symbols, please ensure that you assign them correctly before using them; otherwise, you will generate the Runtime Error (dump) GETWA_NOT_ASSIGNED. This can happen in the following situations (Figure 21 and Coding 47).

```
Error analysis
    An attempt was made to access a field symbol that has not been assigned
    yet (data segment number "-1").

    That error occurs if
    - a typed field symbol is addressed before it has been set with ASSIGN,
    or
    - a field symbol is addressed that points to a row in an internal table
    that has been deleted, or
    - a field symbol is addressed that was previously reset using UNASSIGN,
    or that pointed to a local field that no longer exists, or
    - a parameter of a global function interface is address, although the
    corresponding function module is not active (is not in the list of
    active calls). The list of active calls can be taken from this short
    dump.
```

Figure 21 - Error Analysis of the GETWA_NOT_ASSIGNED Runtime Error

```
    CLEAR <source_fields>.

    LOOP AT source_package ASSIGNING <source_fields>.
```

Coding 47 – Possible Root Cause for the Runtime Error from Coding 45

Performance Analysis and Debugging

If you are following the rules established in the other chapters, you should experience no performance issues in your transformations. However, it sometimes happens that although you entered good coding the logic of the coding does not yield a performant execution. First, we must ask the question: "What is a good performance?" The answer is not that always obvious, especially since HANA has become available as a database. Ten years ago I would have said that one hour for 1 million records was quite OK. Today you can handle 10 million records in less than 10 minutes if you are using the full power of the HANA calculation. What a difference!

If you are using the DTP monitor for analyzing performance, you can see the steps where the coding bottleneck is happening (Figure 22).

15.02.2016 15:07:52	10 Minutes 31 Sec.
:cords 5.02.2016 15:07:52	1 Sec.
15.02.2016 15:07:53	
15.02.2016 15:07:53	2 Sec.
15.02.2016 15:07:55	9 Minutes 59 Sec.
15.02.2016 15:17:54	11 Sec.
15.02.2016 15:18:05	19 Sec.

Figure 22 - DTP Monitor to Analyze Performance Issues

Do not spend time analyzing a transformation that takes only 2 seconds. However, taking approximately 10 min for a 10,000-record case should be a reasonable basis for an analysis.

First, a break in the coding is required to start the debugger. While you can use the predefined Breakpoints from SAP, in my experience, they are always in the wrong place.

Do not use hard-coded BREAKPONITS or personal BREAKPOINTS in your coding. I can almost guarantee that you will forget to take them out before transporting

Since it is not possible to set a breakpoint when displaying the start routine in the transformation, you must use the MENU: EXTRA – Display-generated program. Then, use the spyglass to find the point in your coding where you want to set the breakpoint. For example, use SOURCE_PACKAGE for searching.

Now, set a BREAKPOINT (temporarily) by clicking on the left-hand margin (Figure 23).

```
2822     LOOP AT SOURCE_PACKAGE ASSIGNING <source_fields>.
```

Figure 23 – Session Breakpoint in the Coding

If you are executing your DTP in a serial in debugging mode, the execution will stop at this point and you can then look at your internal tables, variables, or execution times for the function module. Use F5 for single steps and F6 if you want to see the full execution of the function module. One very nice aspect is to set the BREAKPONIT again by clicking on the left margin and then executing the coding to this point with the F8 button.

You can also start the debugging mode if you are in the monitor of a DTP by clicking the **Debugging** button on the menu bar.

Database vs. ABAP Runtime

If you have no success with the debugging technique, you can also use the trace functionality inside SAP to analyze the root cause of the bad performance. First, please ask yourself—did the bad performance originate in the database access or during execution of the ABAP coding? Since we are working with HANA, database access is not generally the issue if the SELECT statement is well programed.

For a runtime analysis, you can use the transaction SE30, but you may also have to ask about the basis for access in quality or test systems. This is a very powerful tool for analyzing the performance of a given code or the execution of an ABAP. Also, in the debugger, please check the possibility for analyzing the used or allocated memory of internal tables and other objects (Figure 24).

St	Variable Name	Variable Type	Bound Used Memory [Bytes]
	SOURCE_PACKAGE	Standard Table[863x106(1276)]	1.102.196

Figure 24 – Memory Analysis for Run-time Objects

If you are working with large objects, this value can give you a hint during the execution of your transformation.

***

ABAP and HANA

There is a mismatch between traditional ABAP programming and the powerful HANA in the in-memory solution. The main reason for this is that the traditional ABAP works on the application server side and HANA and SQL script languages work on the database side. The appropriate language, now universally used, is "push the code down to the database." So, for our ABAP work and especially when handling millions of records, we must either move our logic from internal LOOPs and internal tables to powerful select statements, or we must set up a HANA expert rule directly in SQL script. A third option is to use the old standby "Formula" function instead of routines, because this is supported for possible HANA processing (Figure 25). Beware—some formula aspects are not HANA-supported.

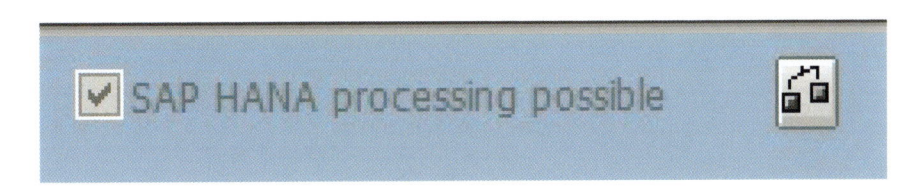

Figure 25 – SAP HANA Flag Inside a Transformation

The following rules are supported by HANA processing Figure 26

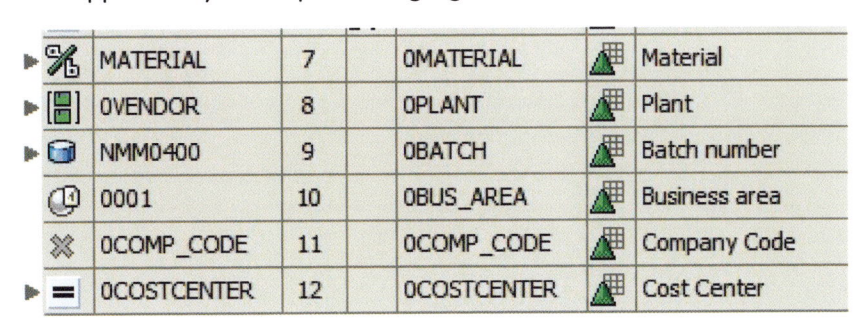

Figure 26 – Possible Rule Types for HANA processing

Because of this issue, if you want to use the full power of the HANA internal processing, you must consider your data model and adjust it so that you can use HANA-supported rule types.

If you have to use other functions for historical or other reasons, you can also write an SAP HANA script transformation, and you will find the entry point for this at the menu bar Edit -> Routines -> SAP HANA Expert Script. Be very careful because, by doing so, the following message appears (Figure 27).

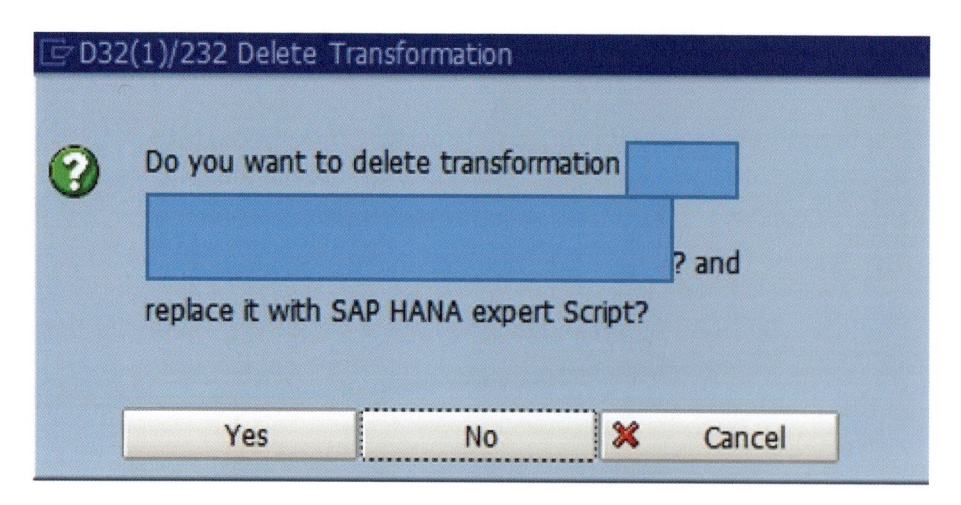

Figure 27– Warning during creation of SAP HANA Expert Script

For security reasons, it is better to create a copy of your old transformation and then test the new HANA. Otherwise your previous work will be deleted.

To learn "How to write SAP HANA SQL Scripts," please refer to the Internet or wait for one of my next e-books.

All about SELECT in SAP BW for HANA

There are a few important points you must be aware of if you are working with HANA as a data base and you want to have performant SELECT statements. SAP Notes cover these points (2000002 - FAQ: SAP HANA SQL Optimization).

In general, every SELECT statement you type in your coding will be translated by the database optimizer into a SQL statement for the database. There are several options in the system for seeing this SQL statement and a so-called execution plan for doing so. To read and understand both of these, you must learn some basic things about the SQL language, which is beyond the scope of this brief guide. To become a better ABAP developer, at the very least, you must read the SAP HELP for the SELECT statement.

To achieve good performance in SELECT and in your ABAP coding overall, always keep the following points in mind:

1) During coding
   a. Know how much data you want to handle.
   b. Become knowledgeable about the key of the database table and use it, if possible, in the same order in your WHERE clause.
   c. If you are using the FOR ALL ENTRIES addition, which is a powerful statement, prior to doing so, check whether the used internal table for selection is empty. (Full table scan, possible!!!)
   d. If you expect a high number of records in your source database table, use a database hint to optimize this, but do not use this hint for everything! Also, if you are not using it in the correct way, your performance will decrease. See the section 'Using the Database Hint'.
   e. To merge two or more different tables easily, you can create a data base view in the dictionary. Sometimes a view is better to test and optimize than a JOIN in the source part of the select statement. In addition, the biggest advantage here is that you can re-use this database view very easily.
2) During testing
   a. When using the Trace Tools, such as Transaction ST05 or the debugger, to determine the specific cause for a long execution time, at best you must do this when nearing the production area, such as in a copy containing a good set of data.
   b. If you have found the bottleneck in your code, ask yourself if you can optimize the situation by reducing:
      i. the amount of data to be handled,
      ii. the length of the data set that must be handled, or
      iii. the time in the ABAP runtime – Application Server (using the powerful HANA instead means that you might be able to use the same logic as in your coding directly by a joined SELECT statement).
3) During optimization
   a. Think about a new logic of your code. Sometimes it is better to re-design the coding or to start from the scratch with more knowledge about "how the ABAP has to work" rather than optimizing small parts of the coding.

b. Keep an eye out for SAP coding, which is produced using the same approach you want to use in your coding. You might find a function module or a static method that you can use in your own coding.
c. Finally, ask a colleague if s/he can work with you to optimize the coding. In my experience, working in a team is one of the best approaches for solving big performance issues.

***

Good Naming

I believe it is important to say something about good naming in ABAP coding so that a code is easy to read and understand for those who did not develop this code. As such, when you begin to create objects in your coding—variables, internal tables or structures—establish a little concept about the style of naming. In addition, please put all the declaration content at the beginning of your coding. Create names that are not too long or complicated.

Also, there is always discussion about the use of hard-coded literals in coding versus the practice of some developers to use constants rather than literals. In my opinion, the most important thing is to make the coding readable and maintainable. In doing so, sometimes it is better to use a literal hard-coded into your ABAP rather than cryptic declared constant values.

***

Using Database HINT for HANA

A data base HINT is an enhancement of the SELECT Statement that tells the database optimizer how to execute the SQL Statement behind the SELECT. For large tables, we recommend using HINT to retrieve data more quickly from the database, especially, if you are using the FOR ALL ENTRIES addition a big chunk of data (Fetch) may also be transferred to the runtime environment.

You can easily enhance your SELECT Statement with the hint in your ABAP coding. You can store the hint in a variable or type it directly into the coding.

Two examples are provided below:

```
SELECT * from (table) INTO TABLE lt_table
     FOR ALL ENTRIES IN lt_material
       WHERE material EQ lt_material-material
       AND objvers EQ 'A'
       AND workcenter NE ''
         %_HINTS HDB '&SUBSTITUTE LITERALS&'.
```

Coding 48 SELECT with hardcoded HINT for HANA data base

Coding 1 shows the hardcoded method of using the HINT Database. In the second option, a function module is used to create or better fill a variable for the HINT. In Coding 2, the function module is used to create a HINT for an FOR ALL ENTRIES (FAE) addition. The export parameters for this coding are as follows:

1) The internal table used for telling the function module which database table to use for the SELECT.
2) The number of Fields to be used in the WHERE clause.
3) The number of lines in the internal table for the FOR ALL ENTRIES addition.

The import parameter is the HINT database for the SELECT Statement.

```
CALL FUNCTION 'RSDU_CREATE_HINT_FAE'
  EXPORTING
    i_t_tablnm       = li_t_tablnm
    i_fae_fields     = 1
    i_fae_lines      = lines( lt_material )
    i_equi_join      = rs_c_true
  IMPORTING
    e_hint           = l_hint1
  EXCEPTIONS
    inherited_error = 1
    OTHERS          = 2.
```

Coding 49 Create HINT using Function Module

Of course, you must first declare the variables, as shown in Coding 3.

```
DATA: l_hint1 TYPE rsdu_hint.
DATA: li_t_tablnm TYPE rsdu_t_tablnm.
```

Coding 50 Variables to be used by the Function Module

In addition, you have to use the variable l_hint1 for the addition in the SELECT statement.

```abap
SELECT * from (table) INTO TABLE lt_table
     FOR ALL ENTRIES IN it_material
       WHERE material EQ it_material-material
       AND objvers EQ 'A'
       AND workcenter NE ''
         %_HINTS HDB l_hint1.
```

Coding 51 Select with a dynamic variable for the HINT

The only remaining question is—what is a large table? Because SAP, as of yet, mentions nothing about this in its documentation, my recommendation is to use HINT for tables greater than one million rows.

***

The Author

Matthias Zinke - Berlin- Germany

Graduate Engineer (Dipl.-Ing.) in Chemistry, University of Applied Sciences, Berlin

Graduate Engineer (Dipl.-Wirt.-Ing.) in Economics, University of Applied Sciences, Berlin

Author of the book: Daten-Remodellierung in SAP NetWeaver BW (SAP PRESS)

Author of the EBook: How to Setup SAP Netweaver 7.3 Trial Version

Since 2001 certificated SAP Consultant

Since 2008 independent Consultant for SAP BW and HANA Applications

Specialist in optimizing of performance for BW Systems and ABAP developer for SAP BW

You can contact the Author at zinke.matthias@gmx.de.

***

Appendix

Table of Figures
Figure 1 - Attributes of an ABAP Report
Figure 2 - User Specific Settings in the SE38
Figure 3 -Settings for the Pretty Printer
Figure 4 - Code Completion I
Figure 5 - Setting for Code Completion II
Figure 6 - Simple Data Flow
Figure 7 - Data Flow with Inbound- and Outbound Info-Source
Figure 8 - Table RSDODSOIOBJ
Figure 9 – Example for SE16 and Table RSDODSOIOBJ
Figure 10 – Output of Transaction SE16
Figure 11 - The Structure Definition of the Monitor Table
Figure 12 – Example for a Look-up Table (DSO)
Figure 13 - Self-update for a DSO Object
Figure 14 - 1:1 Mapping of the Fields
Figure 15 - Technical Rule Group for the 0RECORDMODE
Figure 16 - Routines in DTP Variables
Figure 17 - The Result of the Coding from Coding 21
Figure 18 - Detail View of a Transformation
Figure 19 - Correct Format to Use in a Dictionary Definition.
Figure 20 - Correct use of the SORT and BINARY READ
Figure 21 SAP Help for BINARY SEARCH
Figure 22 - Error Analysis of the GETWA_NOT_ASSIGNED Runtime Error
Figure 23 - DTP Monitor to Analyze Performance Issues
Figure 24 – Session Breakpoint in the Coding
Figure 25 – Memory Analysis for Run-time Objects
Figure 26– Warning during creation of SAP HANA Expert Script

Table of Coding
Coding 1 - The Type-Definition in the Start-Routine
Coding 2 - Type for a second Source Package
Coding 3 - Naming proposal for Local Objects
Coding 4– Definition of the Structure for the Source Package
Coding 5 - Definition (Type) of the Source Package
Coding 6 - The Standard Type Definition in the Start Routine
Coding 7 – The Type for the Monitor Table
Coding 8– Begin of the Global Part for the Start Routine
Coding 9 – The Second part of the Global Part in the Start Routine
Coding 10 – The Main Part for the Coding in the Start Routine
Coding 11 – Using INCLUDE in the Start Routine
Coding 12 – Ssimple LOOP over the Source Package
Coding 13 – Using work-areas instead of pointers
Coding 14 – FREE and CLEAR Objects After or Before Using Them

Coding 15 – The Correct Type for the local objects
Coding 16 – Select Statement for a Lookup from a DSO
Coding 17 - Checking the Content before using FOR ALL ENTRIES
Coding 18 – The READ TABLE Statement in a LOOP
Coding 19 – Manipulating the 0RECORDMODE in a Start Routine
Coding 20 – Adding records in a Start Routine of a DSO
Coding 21 – Example for Date Selection
Coding 22 – Calculation of Number of Days for a given Calendar Day
Coding 23 – Second Step: Calculation of the Average using ASSIGN COMPONENT
Coding 24 - Select statement from table RSKSFIELDNEW
Coding 25 – The right Format for the Date in ABAP Coding
Coding 26 – Using Offset for the date Field
Coding 27 – Using SUBTRACT 1 to Retrieve the Last Day of a Month
Coding 28 – Cleaning out Internal Tables
Coding 29 – DO NOT USE the Obsolete REFRESH Statement
Coding 30 – Table Reference to the Dictionary
Coding 31 – Local Definition of a Table Type
Coding 32 – Table Definition with a Local Type
Coding 33 – Structure Definition with the same Local Type as in Coding 32
Coding 34 – Standard, Sorted and Hash Tables
Coding 35 – Append of an Internal Table
Coding 36 – Completely Moving Content from one Internal Table to Another Internal Table
Coding 37 – READ Statement with the addition KEY only
Coding 38 - LOOP and EXIT for getting data out of an Internal Table
Coding 39 - Correct use of SORT and BINARY READ
Coding 40 - DELETE Statement for an Internal Table
Coding 41 - Return Code of a Successful READ Statement
Coding 42 - First Line of an Internal Table
Coding 43 – Describe Tables to retrieve the Number of Lines
Coding 44 – Variable Selection from a Table
Coding 45 – Generic CALL of a Function module
Coding 46 – Working with ASSIGN COMPONENT OF STRUCTURE TO Statement
Coding 47 – Possible Root Cause for the Runtime Error from Coding 45
Coding 48 SELECT with hardcoded HINT for HANA data base
Coding 49 Create HINT using Function Module
Coding 50 Variables to be used by the Function Module
Coding 51 Select with a dynamic variable for the HINT

Index

0RECORDMODE, 20
ASSIGN COMPONENT OF STRUCTURE, 36
BINARY SEARCH, 33
code completion, 7
data base HINT, 43
date, 28
generic ABAP, 35
Hash Table, 31
Info-Sources, 9
internal tables, 30
lookups, 18
loop, 16
monthly average, 25
offset technique, 28
SE80, 6
start routine, 12

Printed in Poland
by Amazon Fulfillment
Poland Sp. z o.o., Wrocław